WORLD'S BIGGEST ANIMALS

OSTRICH COLOSSAL BIRD

NATALIE HUMPHREY

PowerKiDS press

Published in 2024 by The Rosen Publishing Group, Inc.
2544 Clinton Street, Buffalo, NY 14224

First Edition

Editor: Natalie Humphrey
Book Design: Tanya Dellaccio Keeney

Photo Credits: Cover, p. 1 Cezary Wojtkowski/Shutterstock.com; p. 5 semue85/Shutterstock.com; p. 7 vectorlight/Shutterstock.com; p. 9 Sergei25/Shutterstock.com; p. 11 jo Crebbin/Shutterstock.com; p. 13 Villiers Steyn/Shutterstock.com; p. 15 EcoPrint/Shutterstock.com; p. 17 iModDesign/Shutterstock.com; p. 19 Goldilock Project/Shutterstock.com; p. 21 Dominique de La Croix/Shutterstock.com.

Library of Congress Cataloging-in-Publication Data

Names: Humphrey, Natalie, author.
Title: Ostrich : colossal bird / Natalie Humphrey.
Description: Buffalo, New York : PowerKids Press, [2024] | Series: World's biggest animals | Includes bibliographical references and index.
Identifiers: LCCN 2022055581 | ISBN 9781642825398 (paperback) | ISBN 9781642825404 (library binding) | ISBN 9781642825411 (ebook)
Subjects: LCSH: Ostriches–Juvenile literature.
Classification: LCC QL696.S9 H86 2024 | DDC 598.5/24–dc23/eng/20221208
LC record available at https://lccn.loc.gov/2022055581

Manufactured in the United States of America

CPSIA Compliance Information: Batch #CSPK24. For Further Information contact Rosen Publishing at 1-800-237-9932.

CONTENTS

One Big Bird

What's the biggest bird in the world? The ostrich! This giant bird is known for being the largest around. Like the other birds in its family, the ostrich can't fly. But ostriches don't need their wings to get around. They can run fast!

Where Do They Live?

Ostriches live in the African **savannas**, deserts, and **grasslands**. There are two **species** of ostriches: the common ostrich and the Somali ostrich. Both kinds of ostriches have long legs and a neck with black and white feathers. Somali ostriches have a long blue neck that makes them stand out!

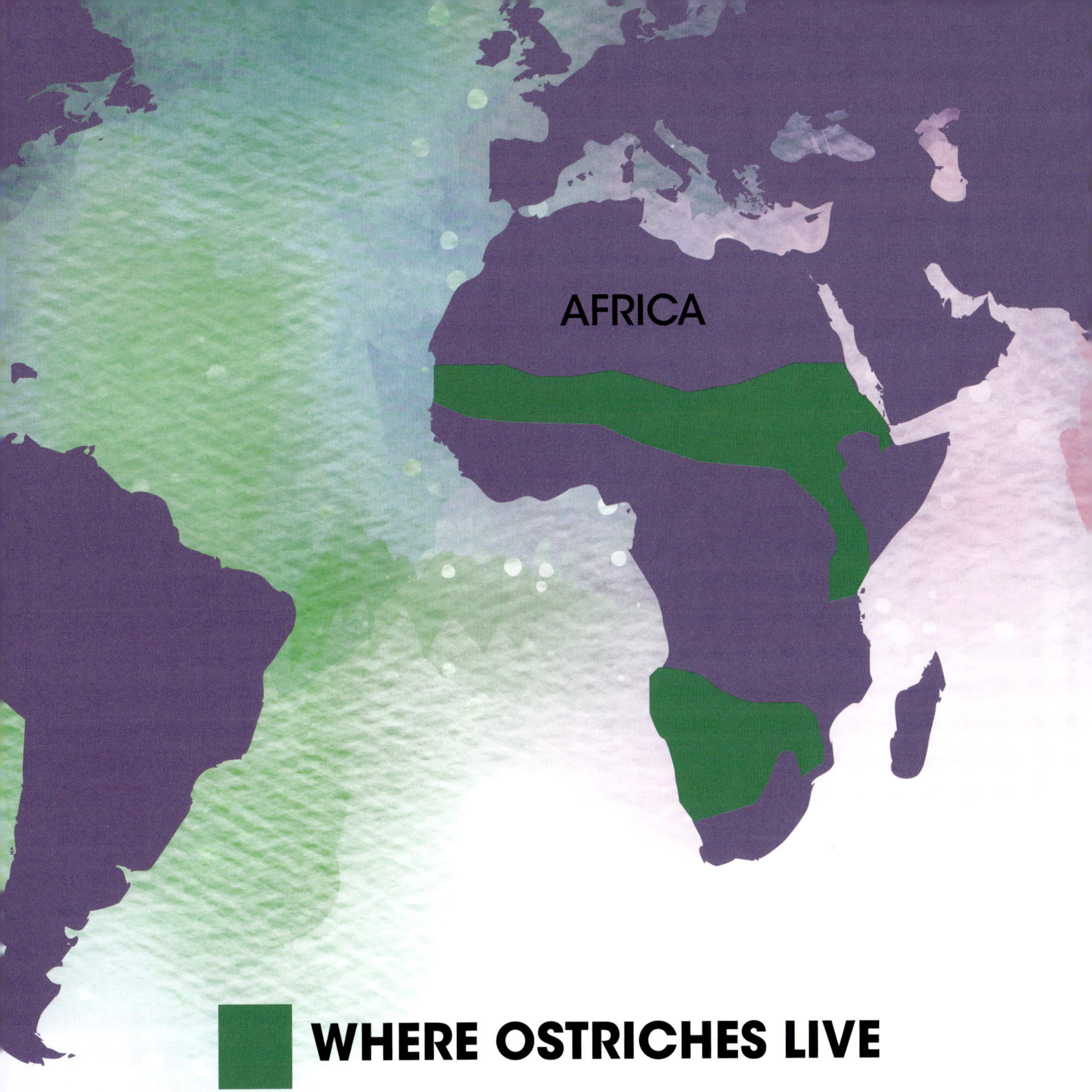
AFRICA
WHERE OSTRICHES LIVE

How Big Are They?

Ostriches are giant birds! A male ostrich can be 9 feet (2.7 m) tall and weigh about 287 pounds (130.2 kg). A female ostrich can be 5.7 to 6.2 feet (1.7 to 1.9 m) tall and weigh 198 to 240 pounds (89.8 to 108.9 kg).

Running Fast

Ostriches can't fly. Instead, ostriches run! When an ostrich is in trouble, it can run over 40 miles (64.4 km) per hour. Ostriches can run long distances without stopping. Just one **stride** from an ostrich can cover up to 16 feet (4.9 m)!

Fact or Fiction?

Does an ostrich really stick its head in the ground when it's scared? No! This story is a **myth**. When an ostrich faces an enemy, it would rather use its speed to run away. If it can't get away, an ostrich fights!

Powerful Kicks

Ostriches don't just use their legs for running. They use them for **protection** too! When an ostrich is in danger and can't run away, it will use its long legs to protect itself. Ostrich legs are so strong that one kick can kill a lion!

Time for Dinner

Ostriches eat grass, insects, and **rodents**. Ostriches usually eat between three and four pounds (1.4 to 1.8 kg) of food a day. Ostriches eat whenever they can, but they don't always need to drink water. An ostrich can go about two weeks without drinking!

An Ostrich Herd

Ostriches live in herds. An ostrich herd may have about 10 ostriches, but some have more. Herds are usually made up of one male ostrich and many female ostriches. The alpha female, or female in charge, takes care of all the eggs from the herd.

Giant Eggs

Even ostrich eggs are huge! Ostrich eggs can be up to 6 inches (15.2 cm) long and weigh 3 pounds (1.4 kg). That's as heavy as 24 chicken eggs. When they **hatch**, baby ostriches are about 10 inches (25 cm) tall.

GLOSSARY

grassland: Land on which grass is the main kind of plant life.

hatch: To break open or come out of.

myth: A story that was told by a people to explain a practice, belief, or natural event.

protection: The act of keeping something safe from harm.

rodent: A small, furry animal with large front teeth, such as a mouse or rat.

savanna: A grassland with scattered patches of trees.

species: A group of plants or animals that are all the same kind.

stride: A long step.

FOR MORE INFORMATION

BOOKS

Emminizer, Theresa. *Awesome Ostriches*. Buffalo, NY: PowerKids Press, 2021.

Rice, Jamie. *Emu or Ostrich?* Minneapolis, MN: Jump! Inc., 2023.

WEBSITES

National Geographic Kids
kids.nationalgeographic.com/animals/birds/facts/ostrich
Check out maps, fun facts, and photos of ostriches.

San Diego Zoo
animals.sandiegozoo.org/animals/ostrich
Learn more fun facts about how ostriches live in the wild.

INDEX